Francielly Carvalho Meneses de Almeida
Neyla Cristiane Rodrigues de Oliveira

# Illustrated Guide to Zoology: a didactic resource for teaching

Francielly Carvalho Meneses de Almeida
Neyla Cristiane Rodrigues de Oliveira

# Illustrated Guide to Zoology: a didactic resource for teaching

## Animals from the Prof. Camillo Filho School

ScienciaScripts

**Imprint**



Cover image: www.ingimage.com

This book is a translation from the original published under ISBN 978-613-9-65814-5.

Publisher:
Sciencia Scripts
is a trademark of
Dodo Books Indian Ocean Ltd. and OmniScriptum S.R.L publishing group

120 High Road, East Finchley, London, N2 9ED, United Kingdom
Str. Armeneasca 28/1, office 1, Chisinau MD-2012, Republic of Moldova, Europe
Printed at: see last page
**ISBN: 978-620-7-80025-4**

# SUMMARY

Dedication ........ 2
PRESENTATION ........ 3
INTRODUCTION TO INVERTEBRATES ........ 4
INTRODUCTION TO VERTEBRATES ........ 31
REFERENCES ........ 48

# I DEDICATE

To the students of 7th grade A (2013), who contributed to the construction of this guide.

## PRESENTATION

This guide is an integral part of the Course Conclusion Work entitled "Guia ilustrado: recurso didâtico para o ensino de zoologia" (Illustrated guide: didactic resource for teaching zoology), the result of field research carried out at the Professor Camillo Filho Municipal School, located in the Esplanada do Uruguai neighborhood, Teresina-PI.

The animals found in the school grounds were photographed and recorded with the help of the 7th grade A students from the class of 2013. All the material collected was identified and transformed into this teaching resource.

The guide shows how rich the school's biodiversity is and how it can be used to arouse students' interest in zoology lessons. The animals that make up this material have been divided into two groups, invertebrates and vertebrates. Within each group they are organized in alphabetical order of their popular names. Each animal in the guide contains taxonomic information, characteristics, habitat, distribution, behavior, food, reproduction, habits and an alternative topic (some animals contain the risks they can pose to humans or their economic importance), as well as a table of curiosities entitled Did you know? At the end of the guide there is a fact sheet entitled: What's the difference? There's also a box called Navegando (Navigating) - with links to websites dealing with zoology and a glossary.

# INTRODUCTION TO INVERTEBRATES

nvertebrate means an animal that does not have a vertebral column or a skull. They are animals that have soft bodies, however there are some, such as arthropods, that are known for having a calcareous exoskeleton, associated with the functions of the internal skeleton of vertebrates, with the aim of support, greater ease of movement as well as protection.

Invertebrates are divided into the following phyla: Porifera, Cnidaria, Platyhelminthes, Nematoda, Mollusca, Annelida, Arthropoda and Echinodermata. They can inhabit any environment and have a varied diet. Depending on the complexity of the phylum, reproduction can vary from a simple asexual form to the most complex forms of sexual reproduction.

The systems of these animals will vary according to the position of the phylum in the cladogram.

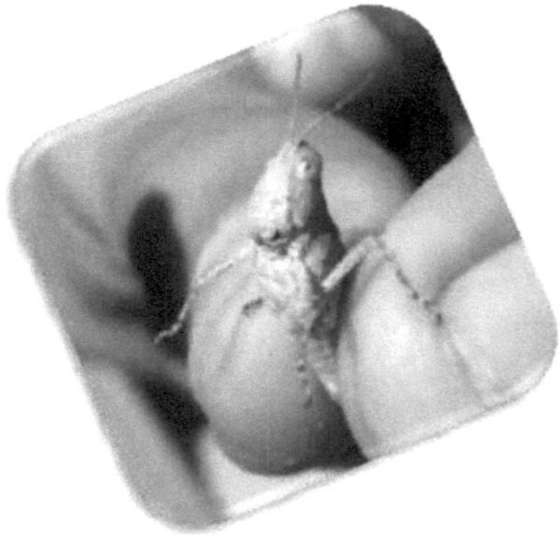

Main references used to support the information contained in this section:

D'Araùjo; Silva (1968).

Gonzaga; Japyassù and Santos (2007).

Leao (2011).

Lima (1955).

Pereira (2012).

Ruppert; Fox and Barnes (2005).

SILVER SPIDER

Photo: Almeida, F. C. M. 2013.

KINGDOM: Animmalia

phylum: arthropoda

CLASS: Arachinida

ORDER: Araneae

FAMiLY: Araneidae

GENRE: *Argiope*

SPECIES: *Argiope argentata*

POPULAR NAME: Silver Spider, Garden Spider or Silver Argiope.

**Characteristics**: *Argiope argentata* is a showy-looking spider with a marked sexual dimorphism; while females reach 18-22 mm (body length without appendages), males only reach 6-8 mm. Females have a large abdomen, visibly flattened dorsoventrally, with three or four pairs of lateral lobes and a posterior terminal prominence; the coloration is variable, and although it can be yellow, white, black and silver, it can appear in other colors. The carapace is completely covered in small hairs. Males have legs and carapace similar to the female, but without visible lobes on the abdomen; the latter is small, almost fusiform.

**Habitat and distribution:** dunes and sand dunes, natural forests, scrubland, orchards and olive groves, meadows, disturbed areas, wastelands or ruderal areas, areas of tall grass, as well as sparse scrubland, in which it builds its web.

**Behavior:** the web is orbicular, large, slightly inclined and with a thickening (stablimentum) typically in the shape of a zigzag, which varies between individuals. The spider is in the center, with its legs grouped two by two. Their webs are built in foliage.

**Food: It** feeds exclusively on insects. Its web is highly resistant and elastic, and is capable of catching a wide variety of prey. It builds large radial webs (they can reach 1m in diameter), usually inclined, at a very variable height from the ground, from a few centimeters to 2m, among dry grasses, shrubs or, more rarely, trees. The spider always remains in the central area of the web, in a characteristic position, bringing its legs together two by two in the shape of a cross. When standing on the web, it mimics the ground as its ventral face is usually turned upwards. The web is very resistant and mainly catches grasshoppers and butterflies. *Argyrodes argyrodes* is often present in the web of this species.

**Reproduction:** the mating ritual begins when the male approaches the female, making spasmodic movements with his forelegs. The female, upon realizing his presence, is passive, raising her body a little. Meanwhile, the male walks under the female, adopting a "belly to belly" position. He then inserts the palps (male copulatory organs) alternately into the female's epigynum. After mating, the male often dies - as far as we know, this is not necessarily the fault of the female. The eggs are laid in thick silk oothecae which the females attach to the grass. The mother never witnesses the birth of the young, as she dies first. Thus, there is no parental care.

**Habits:** these spiders reach adulthood in summer and fall. They remain active during the day, building or renewing their webs before dawn.

**Importance:** although they are not aggressive, this species is often referred to as a "very painful stinging species" but is considered harmless because it does not cause serious or permanent damage.

## DID YOU KNOW?

Although they are not aggressive, this species is often referred to as a "very painful stinging species", but is considered harmless because it does not cause serious or permanent damage. A curious habit is to swing their web when they sense movement nearby or when they are touched. This behavior can have a double function: in hunting, it increases the efficiency of the web, both by increasing the contact surface between it and the prey, and as a defense, as they become more difficult to locate and capture.

**FLYCATCHER SPIDER**

Photo: Almeida, F. C. M. 2013.

KINGDOM: Animmalia

phylum: arthropoda subphylum: chelicerata class: arachnida order: araneae infraorder: araneomorphae family: salticidae subfamily: plexippinae tribe: hylini genus: philaeus

CIENTIFIC NAME: *Philaeus sp*

POPULAR NAME: flycatcher, jumping spider.

Characteristics: this is the family with the most species. Small spiders, smaller than 15 mm. Spiders from the Salticidae family do not build webs. They have 8 eyes and the sharpest vision of all spiders, which is important for their predatory ability. The arrangement of the eyes, four large ones on the face and four smaller ones on the top of the head, distinguishes them from other spiders. At the end of each leg there are hairs, each of which is then divided into hundreds of tiny hairs, each end with an "end foot". These thousands of tiny feet allow them to climb up and over almost any terrain, even glass.

**Habitat and distribution: they** like to live in woods, forests, fields, rural areas, but are commonly found in urban areas. Found throughout Brazil, this family has an almost worldwide distribution.

**Behavior:** For humans, it is considered curious, it can even keep its distance, but it can also jump on a person, in a curious way (it is usually interested in everything that comes near it), not out of aggression.

**Food:** carnivorous, feeding on insects and other spiders.

**Habits:** they are diurnal and active hunters. They are known for their leaps towards their prey. They don't build webs, but make small silk shelters under leaves or bark and actively hunt their prey during the day, tracking and then jumping on their victims.

**Importance:** Jumping spider bites are not considered harmful to humans, the venom rarely results in more serious effects than skin irritation (swelling, redness around the skin). However, the initial effect of the bite can be painful.

## DID YOU KNOW?

Spiders from this family are capable of jumping 100 times their size.

They have a hydraulic system that extends to their limbs, altering the pressure of the body fluid (hemolymph) inside them. This allows them to jump without having large muscular hind legs, like grasshoppers.

Some people may have an allergic reaction to the venom of a jumping spider. Symptoms of allergic reactions include dizziness, nausea, fever, vomiting and open sores.

## PUPPET

Photo: Duarte, J. Jr. 2012.

KINGDOM: Animmalia

phylum: arthropoda

SUBPHYLA: Hexapoda

CLASS: Insecta

ORDER: Phasmatodea

SUPERFAMILY: Phasmatoidea

POPULAR NAME: "bicho-pau", "mané magro", "bicho-palha", "chico magro".

**Characteristics:** there are more than 500 species of this insect. There are stick insects with the most different shapes, commonly in Brazil we find lichen-like shapes on tree trunks. These insects have a stick-like body and may or may not be winged. The wingless forms move slowly. The winged ones fly poorly, using the wing more as a parachute. They are sexually dimorphic, the male being smaller than the female.

**Habitat and distribution:** lives among plants in rural and urban areas. It is found throughout Brazil and in tropical and subtropical regions all over the world.

**Behavior:** harmless, it prefers nightlife and is capable of spending hours completely paralyzed to escape predators, as well as having the ability to camouflage itself.

**Food:** they are herbivores, feeding on leaves and shoots.

**Reproduction:** the woodworm multiplies through oviparous reproduction; its eggs are dropped by the female in various directions. These eggs are like germs, casually placed on the ground by the female specimens. Development is slow, and when they hatch they are called nymphs and have a similar physical shape to the adult. Some species perform parthenogenesis - the females lay eggs without mating and these eggs give rise to offspring that have exactly the same genetic heritage as the mother, and can be considered maternal clones.

**Habits:** the mother does not care for her offspring. This animal moves by means of oscillations, which is why it is often mistaken for a leaf swaying in the wind.

**Importance:** the woodworm feeds on herbs, foliage and shoots, which is why in urban areas it can be seen on guava trees or pitangue trees. However, it never multiplies enough to ruin agriculture.

## DID YOU KNOW?

In Brazil, the largest specimens belong to the *Bactridium grande* species, which reaches almost 27 centimetres.

As well as moving slowly to protect themselves from predators, these animals emit a disgusting milky fluid to defend themselves. Other species have the power to regenerate lost limbs, as well as the gift of reviving eliminated limbs.

The development of the embryo is slow, taking between 100 and 150 days for the egg to hatch. In some species, this can take up to two years.

# SMALL TOMATO BORER

Photo: Almeida, F. C. M. 2013.

KINGDOM: Animmalia

phylum: arthropoda

CLASE: Insecta

ORDER: Lepidoptera

FAMiLY: Crambidae

SUBFAMILY: Pyraustinae

TRIBE: Spilomelini

GÈNERO: *Neoleucinodes*

SPECIES: *Neoleucinodes elegantalis* POPULAR NAME: small fruit borer, small tomato borer.

**Characteristics:** a moth with a wingspan of around 25 mm and a white color. The wings are transparent, the front wings have a brick-colored spot and the back wings have small brown spots. The adult's lifespan is around 6 days. They are sexually dimorphic, with the female being larger than the male. The female has a large abdomen with a truncated end, while the male has a slender abdomen with a sharp end covered in a tuft of brush-like scales.

**Habitat and distribution:** intertropical region, where solanaceae are grown.

**Behavior:** the moth lays its eggs on the surface of the fruit. When they hatch, the caterpillars take a few minutes to begin piercing the fruit. The larva remains feeding inside the fruit and when it reaches the pupal stage, it burrows into the soil, leaving a wound that will allow pathogenic microorganisms to enter the fruit.

Food: they feed on solanaceous plants, particularly tomatoes, eggplants and peppers.

Reproduction: the females produce a sex pheromone to attract the male. Adults remain motionless throughout the day, with their wings outstretched to the sides and their abdomen recurved. The first sign of activity is the extension of the abdomen, with slow movement and short flights. After this period, the active female walks along the surface of the fruit with her abdomen recurved and then distends it for oviposition. Most males only copulate with one female.

Habits: the moth has crepuscular and nocturnal habits.

Importance: it is considered a key pest in tomato cultivation because it damages the reproductive parts.

## DID YOU KNOW?

In Brazil, the small borer is responsible for an average loss of 45% of national production, reaching 100% in some regions.

This pest is combated through biological control, one of the most widely used being *Trichogramma, a* group of entomophagous insects, which is an egg parasitoid.

## CRISOPA

Photo: Reyes, D. M. 2012.

KINGDOM: Animmalia

phylum: arthropoda

CLASS: Insecta

ORDER: Neuroptera

FAMiLY: Chrysopidae

GÈNERO: *Chrysoperla*

SPECIES: *Chrysoperla carnea*

POPULAR NAME: Crisopa.

**Characteristics:** they are delicate insects with a wingspan of between 6 and 65 mm, their bodies are usually green and shiny, their eyes are usually golden and their wings are translucent with greenish veins. They are holometabolous animals (they have a larval stage) and are the most important predators of the order Neuroptera. They are currently considered important in pest control.

**Habitat and distribution: all over** the world, except Australia, it is found in undergrowth, shrubs and trees.

**Behavior: there** is a color change, from green to gold in autumn, and in the opposite direction in spring. Some lacewings flee when they hear the ultrasound of bats. When fleeing, they close their wings and head for the ground.

**Feeding:** Neuroptera are avid flea eaters. During the larval stage, known as the predatory stage, they feed on mites and lepidopteran eggs, in addition to cannibalism; in this stage they are active predators, have a specialized mouth and extraintestinal digestion. In the adult stage, they feed on honey, dew, pollen, fungi and algae.

**Reproduction:** during the mating phase, the abdomen vibrates to attract a mate. The eggs are deposited imago, either in groups or alone, have 3 larval stages and 500 to 600 eggs are laid per female. Reproduction of these animals averages 2-3 generations per year.

**Habits: they** have a crepuscular and nocturnal habit, often seen around lights at night, remaining hidden during the day under wet grass. The lacewings try to carry out their biological activities during this period.

**Importance:** they are important in biological control, and can be applied in greenhouses and open ground. The larvae are voracious and adaptable predators.

## DID YOU KNOW?

Depending on the species, some chrysopids eat only 150 prey during their lifetime, while in other cases 100 aphids can be eaten in a single week. Thus, in several countries, millions of voracious lacewings are bred for sale as predatory agents for the biological control of insects, aphids, mites, whiteflies and mealybugs in agriculture and gardens. Eggs are sold, with larvae hatching in the field.

## CYANOPEPLA

Photo: Almeida, F. C. M. 2013.

KINGDOM: Animmalia

phylum: arthopoda

CLASS: Insecta

ORDER: Lepidoptera

SUPERFAMILY: Noctuoidea

FAMiLY: Arctiidae

SUBFAMILY: Ctenuchinae

GENRE: *Cyanopepla*

SPECIES: *Cyanopepla jucunda*

Characteristics: forewings, usually triangular in outline, with a dark color and showy stripes, ciliated antennae.

Habitat and distribution: southern grassland, savannah and Atlantic forest. Found in tropical regions, mainly in South America.

Reproduction: the hair on the antenna is used by the males to detect the pheromone of the females.

Habits: diurnal or semi-diurnal.

**EMBUA**

Photo: Almeida, F. C. M. 2013.

KINGDOM: Animmalia

phylum: arthropoda

CLASS: Diplopoda

ORDER: Polydesmida

FAMILY: Paradoxosomatidae

POPULAR NAME: embuà, gangugi, gouro or mil pés.

**Characteristics:** the body is flattened, with black dorsal plates. Há a bright yellow spot on both sides of each body segment. They have one pair of antennae, simple eyes and two pairs of locomotive legs per segment.

**Habitat and distribution:** appearance is always during the warmest and rainiest periods.

**Behavior:** when it feels harassed, it curls up to protect itself. Some species release a colored liquid for defense. They walk slowly, as they generally feed on dead organic matter, both animal and vegetable.

**Food:** they are herbivores (vegetables) and detritivores (decomposing matter).

**Reproduction:** Reproduction is sexual. All diplopods are oviparous, the eggs are laid in the soil.

**Habits: they** live under leaves, stones and logs, but sometimes they even appear inside some houses. They always like damp places.

**Importance: important** in agriculture, as some are detritivores and others feed on seeds, causing some losses.

## DID YOU KNOW?

If you happen to handle the animal, it can release a dye. This paint will remain on the person's hand and fingers for a day or two, but it causes absolutely nothing - no allergies, no pain, no burns, nothing.

**HOPE**

Photo: Almeida, F. C. M. 2013.

KINGDOM: Animmalia

phylum: arthropoda

CLASS: Insecta

ORDER: Orthoptera

SUBORDER: Ensifera

FAMiLY: Tettigoniidae

GENUS: *Tettigonia*

SPECIES: *Tettiigonia Viridissima*

Characteristics: Hope, a species of grasshopper belonging to the genus *Tettigonia, is* characterized by its broad forewings. Generally green in color, although many other species come in different colors, it has a long, curved ovipositor tube (an organ used by female arthropods to lay their eggs), resembling an oriental sword, as well as very long, thin antennae, unlike common grasshoppers, which have shorter ones.

Habitat and distribution: open areas with a hot, dry climate.

Behavior: its adult life is short, as it only lasts a little longer than the three months of summer. With the approach of winter and the consequent record low temperatures, they can't resist the cold and end up dying.

Food: they are carnivorous. Unlike the grasshopper, the hope feeds on pests and not on crops.

Reproduction: before dying, the female lays her eggs in the ground during the fall. These resist the unfavorable climate and hatch in spring, when the nymphs begin their growth process.

Habits: they use their antennae to find prey. They have a great capacity for mimicry. They have a nocturnal habit of life, producing a sound similar to that of cicadas.

Importance: pest control.

**SPIDER CAT**

Photo: Almeida, F. C. M. 2013.

KINGDOM: Animalia

phylum: arthropoda

CLASS: Insecta

ORDER: Lepidoptera

FAMiLY: Limacodidae

GENRE: *Phobetron*

SPECIES: *Phobetron hipparchia*

POPULAR NAME: spider caterpillar.

Characteristic: exotic in appearance, it is a moth caterpillar of the *Phobetron* genus. With a body covered in bristles, this species is sexually dimorphic as an adult, and its larva can be yellow, gray or brown, regardless of its sex. The venomous bristles are restricted to small clusters located on the anterior and posterior regions of the body.

**Habitat and distribution:** found in Mexico, Panama, Ecuador, Colombia, Venezuela, the Guianas, Brazil and Argentina.

**Behavior: the** characteristic behavior of this caterpillar is the detachment of these tentacles in dangerous situations.

**Feeding:** It feeds by scraping the surface of leaves, ending up building a silk cocoon at the end of its development, where it impales itself.

**Reproduction:** when moths mate, they lay their eggs on a plant on which the caterpillars will feed. At this stage they change their skin (ecdysis) several times until they form a silk cocoon in which they will undergo metamorphosis - thus starting the life cycle all over again.

**Habits:** its adult is a moth, with a nocturnal habit.

**Importance:** It is considered a pest of some plants, such as cashew, oak, guava, palm and rose trees, among others.

## DID YOU KNOW?

Moth caterpillars can cause accidents, known as *erucism* (poisoning caused by a moth larva). Contact with this type of caterpillar can cause burning and pain.

## PRAISE GOD

Photo: Almeida, F. C. M. 2013.

KINGDOM : Animmalia

phylum: arthropoda

CLASS: Insecta

ORDER: Mantodea

FAMiLY: Stagmatopterinae

GENRE: *Oxyopsis*

SPECIES: *Oxyopsis sp*

POPULAR NAME: Mantis, praying mantis.

Characteristics: they often appear to be praying, but when they raise their heads and put their front paws together, they are actually waiting for their food. They are slender, usually measuring 5 centimeters. They are sexually dimorphic, with the female being larger than the male. The praying mantis is usually brown or green, which helps it to blend in with its surroundings and escape predators. It has a triangular-shaped head with large eyes located at the top. Its neck is very flexible, so it can look over its shoulder to both the right and the left.

Habitat and distribution: they live in warm regions, especially in the tropics.

Behavior: all species of this order are carnivorous and their diet consists of other arthropods, such as flies, butterflies and grasshoppers, but they can also feed on small vertebrates, especially lizards and micromammals.

**Food:** they are predators, in some situations they practice cannibalism.

**Reproduction:** as the females are larger, they often end up eating their partners, devouring the male's head during mating. In this way, they gain nutrients and energy for laying eggs. The females lay between 10 and 400 eggs in a hardened capsule (ootheca) which they place on the ground, on a flat surface or wrapped in a leaf. In some species, the females stay close by and protect it from predators. The embryos spend the unfavorable season protected by the ootheca and hatch the following spring in the form of small nymphs, similar to adults but still without wings. Like the adults, the nymphs are fearsome hunters, running or jumping after small insects and spiders.

**Habits:** they have diurnal habits.

## ANT BEDBUG

Photo: Almeida, F. C. M. 2013.

KINGDOM: Animmalia

phylum: arthropoda

CLASS: Insecta

ORDER: Hemiptera

SUBORDER: Heteroptera

FAMiLY: Alydidae

SUBFAMILY: Alydinae

POPULAR NAME: Ant bug, ant.

Characteristics: its length varies from 1 to 2 cm. It has an elongated, thin body and is brown in color. The compound eyes are globular and protruding, as well as having ocelli.

Habitat and distribution: most of the genera in this family are from tropical and subtropical regions. Its occurrence in Brazil is more common in the Southeast and Midwest regions, but in recent years there has been an expansion and it can be found in other regions, including the Northeast.

Behavior: when harassed or when they feel threatened, these animals release a characteristic odor, reminiscent of halitosis.

Food: their main food is seeds, which they pierce with their snout to drink the nutritious liquids contained therein.

Reproduction: the eggs are brown and found in groups on the leaves. There is sexual dimorphism, the male is light brown and the female is darker and has a larger abdomen.

Importance: they are considered pests because they attack crops, especially legumes, whose grains are fully formed. They cause a reduction in the quality of grains and seeds, as they become stained due to the action of opportunistic fungi.

## PIRATE BUG

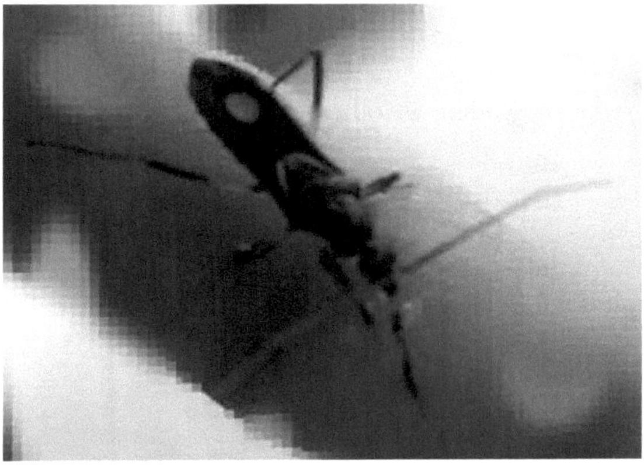

Photo: Almeida, F. C. M. 2013.

KINGDOM: Animmalia

phylum: arthropoda

CLASS: Insecta

ORDER: Hemiptera

FAMiLY Reduviidae

SUBFAMILY: Peiratinae

GENRE: *Rasahus*

SPECIES: *Rosahus hamatus*

POPULAR NAME: pirate bug; false barber.

**Characteristics:** The mouth apparatus (trunk) is very robust. They have an elongated head and a short trunk. Unlike the other representatives of the order, they don't keep their trunk in their mouth.

**Habitat and distribution:** coastal, wet forest, agricultural. South America, Central West Indies and Europe.

**Behavior:** they are predatory killer insects. They eat small animals such as snails, insects, including caterpillars, spiders and other arthropods. To prey on their prey, they attack the animal with their stylus-like proboscis and inoculate venom, which paralyzes the prey, after which they suck out the animal's bodily fluids.

**Feeding:** they are polyphagous, insect predators. As it preys on other insects, it needs an efficient mouth apparatus, precisely so that it can penetrate the "skin" of other insects and suck out their internal liquids.

**Importance:** not considered a pest as they feed on other insects. When they feel threatened they can sting, which is very painful.

## DID YOU KNOW?

They use their proboscis to inoculate a lethal saliva, which liquefies the inside of the prey, then they suck out this body fluid.

## SARA-PAU

Photo: Almeida, F. C. M. 2013.

KINGDOM : Animmalia

phylum: arthropoda

CLASS: Insecta

ORDER: Coleoptera

FAMiLY: Cerambycidae

GENRE: *Retrachydes*

SPECIES: *Retrachydes thoracicus thoracicus (Olivier, 1790)*

POPULAR NAME: saw palmetto, guava borer, sawyers.

**Characteristics: It** has a pair of long antennae (usually longer than the rest of the body), a striking feature of the Cerambycidae family. This beetle cuts the bark of trees, branches and trunks using its strong mandibles (that's where its name comes from). They are sexually dimorphic (the males have antennae that are larger than their bodies). The body of these insects has beautiful markings, which puts the species at risk of extinction, as they attract collectors.

**Habitat and distribution:** representatives of the family can be found on all continents except Antarctica.

**Behavior:** the adults inoculate fungi that will serve as food for the larvae, complementing the consumption of xylem (an energy-poor source), as well as aiding in the degradation of the wood itself, thus facilitating consumption by larvae and adults.

**Feeding:** adults, like other phytophages, are phytophagous, i.e. they are usually found near plants, on flowers, feeding on pollen or eating the flesh of ripe fruit that has already opened; sometimes they are found on the ground.

**Reproduction:** the females lay their eggs on branches, incising them from distance to distance, depositing an egg at the bottom of each incision. However, below the part of the branch on which a number of eggs have been laid, the female cuts it off, sectioning the branch into several increasingly thicker sections. The larva will undergo metamorphosis inside the incision, during which time it builds canals. Subsequently, the adult pierces the bark to the outside, leaving a hole (borer) through which it will exit.

**Habits:** the larvae of these sawflies drill into branches. Their metamorphosis takes place inside the tree's internal chambers.

**Importance:** they are considered pests because their larvae open vast galleries inside the wood.

## DID YOU KNOW?

When the larva hatches, it widens the chamber, the terminal part of the gallery it has pierced to feed, prepares a channel to the inside of the bark, closes the entrance to the chamber with a plug of fibers cut from the stem tissues and, retreating into the chamber, metamorphoses there into a pupa, white or ebony in color. The adult that emerges spends a resting period acquiring its characteristic colors.

## LITTLE GARDEN TAT

Photo: Almeida, F. C. M. 2013.

KINGDOM: Animmalia

phylum: arthropoda

SUBPHYLA: Crustacea

CLASS: Malacostraca

SUPERORDINATE: Peracarida

ORDER: Isopoda

SUBORDER: Oniscidea

POPULAR NAME: beaded animals, garden armadillos, ball armadillos, St. Anthony's pig.

Characteristics: Currently, approximately 120 species are known in Brazil. This is the group with the highest species richness among crustaceans. They are independent of the aquatic environment, as they have an incubator pouch like a marsupium, a flattened body, thick cuticle and structures for gas exchange (pleopodal lungs). They are sexually dimorphic. Females are larger than males.

Habitat and distribution: they generally live in humid environments sheltered from light, found in a variety of habitats, from the coastal zone to deserts.

**Behavior: they** produce aggregation pheromones in the intestinal cells, expelling them along with the feces, so that they can be perceived by sensory structures in the antennae. These pheromones, produced with greater intensity in drier conditions, lead the isopods to form a group and remain in it, reducing the loss of water to the environment. In humid and hot environments, terrestrial isopods use evaporation to maintain their body temperature and cool themselves down.

**Feeding:** they feed at night, preferably on young plants.

**Reproduction:** most species are unisexual and oviposit, incubating their eggs in the cavity formed by the oostegites. When the young break out of these eggs, they are in the post-larva stage, with the absence of the seventh pereiopod, the last pair of legs incompletely developed, a whitish color, a fragile integument and no secondary sexual characteristics. Reproduction is sexual, where the male presses his ventral flank against one side of the female and injects the sperm into one of the female's gonopores using a vibrating pleopod, repeating the process on the other side. Females in their sexual state are recognized by the male through the antenna, probably due to the action of pheromones. Direct development of offspring.

**Habits:** they have defense mechanisms, which consist of: curling up into a ball - which also reduces water loss through evaporation; fleeing; playing dead; clinging intensely to the ground; and releasing repellent secretions.

**Importance:** crustaceans are important for recycling nutrients, accumulating heavy metals and feeding animals, making up a large part of the soil fauna and influencing its dynamics.

## DID YOU KNOW?

That there are intracellular organisms that cause feminization and the appearance of intersex individuals in terrestrial isopods, such as Wolbachia, which alters sexual reproduction strategies and sex ratios in order to guarantee maternal transmission, the only means of infecting offspring.

Some species can cause losses of up to 40% in peppers, up to 70% in tomatoes and up to 80% in bean plants. They can also attack peas and other vegetables.

The isopods' marsupium (incubator pouch) has the function of the initial development of the young, without dependence on external sources (water), with fluid apparently being secreted from the maternal body wall, forming a micro aquarium, and gas exchange probably taking place between the hemolymph and the marsupial fluid.

After 24 hours from birth, the first ecdysis occurs in the little tortoises, which then take place every 3 or 4 weeks until they are about 6 months old, and they can live for 3 to 4 years.

## WASP AMASSA-BARRO

Photo: Almeida, F. C. M. 2013.

KINGDOM: Animmalia

phylum: arthropoda

CLASS: Insecta

ORDER: Hymenoptera

SUBORDER: Apocrita

FAMiLY: Sphecidae

GENRE: *Sceliphron*

SPECIES: *Sceliphron fistularium*

POPULAR NAME: wasp or mud wasp.

**Characteristics:** complete sausage-shaped nests, arranged side by side along a horizontal axis, and found attached to electrical wires and walls. The adults of this species have a striking yellow and black coloration and a narrowing between the thorax and abdomen.

**Habitat and distribution:** it is distributed all over the world, except in the polar regions.

**Behavior:** they are not aggressive towards humans if they are not attacked. They are solitary and use clay to build nests, usually in shady areas, sometimes inside doors and windows;

**Food:** flowers, natural fruit juice and some insects to feed its young.

**Reproduction:** After mating, each female builds her own nest and ensures the complete development of the larvae. Each clay cell contains one egg and females can lay an average of 15 eggs in their lifetime.

**Habits: the** nest is built by the female, who can build one cell a day. The clay is transported from its place of origin to the nest via dozens of trips by the female. They add new cells to the nest one by one, and as the cells are built they are supplied with prey (spiders paralyzed to feed the larvae).

**Importance:** they are considered beneficial due to their control of the spider population and the pollination of various types of plant.

## DID YOU KNOW?

Wasps shape the clay before it dries. When it finds some damp earth, the wasp picks up a pile of clay and tests the humidity while it works. If the clay is too wet, it leaves it to dry, if it's too dry, it moistens it.

Various parasites attack these nests, such as the cuckoo wasps of the Chrysididae family, which lay their eggs in the nests of the *Sceliphron,* and both the larva or egg of the *Sceliphron* and the spider are devoured by the cuckoo wasp larva.

Wasps generally have an elongated abdomen, which gives them great mobility and is very useful when hunting various insects.

The stinger on the tip of the abdomen is active, as wasps usually anaesthetize insects or small spiders with their venom before taking them to their nests.

They prepare a cavity in the nest and place the spiders and other anaesthetized insects there, on which they lay their eggs.

Once the brood cells have been closed, they will never have contact with their offspring again.

## INTRODUCTION TO VERTEBRATES

Vertebrate means an animal with a backbone and a skull.

It is a large division of the animal kingdom, which includes animals with a vertebral column and, generally, two pairs of vertebrae.

of limbs. There are seven living classes of vertebrates: Agnatha, Iamprey and the like; Chondrichthyes, sharks and other cartilaginous fish; Osteichthyes, fish with bones; Amphibia, frogs and other amphibians; Reptilia, reptiles; Aves; and Mammalia, mammals.

Vertebrates live in practically every habitat on Earth. The behaviors of vertebrates are as diverse and complex as their forms. The life of a vertebrate is energetically expensive, and vertebrates obtain the energy they need from the food they eat.

Carnivores feed on other animals and have a wide variety of methods for catching prey. Herbivores eat plants. Herbivorous vertebrates display a set of specializations to cope with the difficulties of feeding on plants.

During reproduction, vertebrates show a surprising array of behaviors associated with mating and reproduction. In general, males court females and the latter take care of the young, but these rules can be reversed.

The vertebrates are represented by the Chordates Phylum, and there are four characteristics that separate the chordates from the others: notochord, dorsal nervous tube, pharyngeal slits and pouches and post-anal tail. These features are always found at some embryonic stage, although they can be modified or disappear at later stages of a chordate's life cycle.

Main references used to support the information contained in this section:

LEÂO, 2011.

POUGH; JANIS and HEISER, 2008.

REIS, 2007.

SOARES, 2009.

## ANU-PRETO

KINGDOM: Animmalia

Phylum: Chordata

CLASS: Birds

ORDER: Cuculiformes

FAMILY: Cuculidae

SUBFAMILY: Crotophaginae

GENUS: *Crotophaga*

SPECIES: *Crotophaga ani*

POPULAR NAME: anu or black anu.

Photo: Almeida, F. C. M. 2013.

**Characteristics:** measures around 36 cm, has completely black plumage, a long tail and a black, high beak. No sexual dimorphism. The smell of its body is strong and characteristic, perceptible to us from several meters away and capable of attracting hematophagous bats and carnivorous animals.

**Habitat and distribution:** fields, pastures and cultivated areas, preferring damp places. It can be found from Florida to Argentina and throughout Brazil.

**Behavior:** very adept at jumping and running between branches, they always live in flocks. They are extremely sociable birds. Weak fliers can barely resist the breeze; any stronger wind will blow them away. The black anu has more than a dozen different voices. Although the females have individual nests, they often team up with one or two couples from their flock to build a collective nest, lay eggs and raise their offspring together, with the cooperation of the males and the grown-up chicks from previous layings.

**Food:** essentially carnivorous, eating grasshoppers, bedbugs, spiders and myriapods. They also prey on hairy and stinging caterpillars, lizards and mice. They fish in shallow water and periodically eat fruit, berries, coconuts and seeds, especially in the dry season when there is a shortage of arthropods.

**Reproduction:** the eggs make up 14% of its body weight and are blue-green in color, covered in a calcareous crust. During mating, the male dances around the female on the ground. Incubation is short, lasting 13 to 16 days. The male usually brings food when he visits the female in the nest. The female lays 4 to 7 eggs.

**Habits: they** like to sunbathe and bathe in the dust, their plumage sometimes becoming heavily tinged with the color of the local soil or with ash and charcoal, especially if they run through wet grass first, as their feathers become sticky. In the morning and after the rains, they land with their wings spread to dry themselves. At night, to keep warm, they usually join other individuals in tight rows or groups.

**Importance:** the black anu is an excellent insect controller, favoring the entire food chain.

## DID YOU KNOW?

That the anu has two alarm chirps: one shout for all the members of the gang to perch in clearly visible spots, examining the situation; and the other, emitted when a caribou approaches, makes the whole group disappear into the undergrowth in an instant.

They amuse themselves by singing softly, in a very varied way, sometimes giving the impression that they are trying to imitate the voice of another bird.

Anus like to be near cattle to catch insects. This happens when the pasture is "trampled" by the cattle, so the insects move and are caught by them.

When there are no cattle in the pasture, the anus perform collective hunts in the field, the flock spreads out on the ground in a semicircle, two or three meters apart. They remain motionless and attentive and when an insect appears, the nearest bird jumps up and catches it.

When they land on the backs of oxen, it is usually to widen their field of vision.

Sometimes they follow tractors that plow the fields, with the same purpose described above.

## IGUANA

KINGDOM: Animmalia

phylum: chordata

SUPERCLASS: Tetrapoda

CLASS: Reptilia

ORDER: Squamata

FAMILY: Iguanidae

GENRE: *Iguana*

SPECIES: *Iguana iguana*

POPULAR NAME: Iguana, chameleon, green iguana, sinimbu.

Photo: Almeida, F. C. M. 2013.

**Characteristics:** only this species of iguana occurs in Brazil. They are herbivorous animals and can reach 10 kilos and 180 centimeters in length, 2/3 of which refers to their tail. The head has numerous scales, which are generally asymmetrical and small. From the nape of the neck to the tail, there is a ridge, and in the throat, a dilatable sac. The nails are strong and pointed and are found on each of the five toes. The life expectancy of these animals is 15 years.

**Habitat and distribution: they** inhabit the tropical areas of the Americas and the subtropical areas from northern Mexico to Paraguay and southwestern Brazil, as well as several Caribbean islands. They can be found both in the Caatinga and in the humid forests of the Amazon.

**Behavior:** they communicate primarily by visual signals. Dominant males display physical characteristics that emphasize their size, such as large heads, opercular scales, ridges, spines, pappus and sometimes orange coloration, cervical and abdominal protuberances, cervical expansions and pappus erection are used to express threats.

**Feeding:** they are completely herbivorous. In some cases, juveniles with insectivorous habits have *been* recorded.

**Breeding:** Breeding is seasonal and associated with the dry season. Males display themselves on branches and defend their territory from November to January, becoming very aggressive towards invading males. The testicles increase in size and produce a marked amount of testosterone. The selection of a female mate is determined by the health of the male. The female produces 14 -76 eggs laid in holes in the ground, which are then covered with leaves.

**Habits:** they are diurnal and arboreal, preferring the tops of trees along rivers, lakes, swamps and mangroves. Iguanas sporadically come down to other trees to protect themselves from possible dangers or to make a nest.

**Importance:** iguanas play a role in the socio-economic development of rural communities, making use of their hides, meat and eggs.

## DID YOU KNOW?

An iguana can stay in a tree from 1 day to several weeks.

These animals spend 90-96% of their time resting and only 1% feeding.

The green iguana changes color to protect itself from predators.

Just like the lizard, the iguana can lose part of its tail to escape from a predator, and then it grows back.

## MORCEGO

KINGDOM: Animmalia

phylum: chordata

CLASS: Mammalia

ORDER: Chiroptera

SUBORDE: Microchiroptera

FAMiLY: Phyllostomidae

POPULAR NAME: bat; andirà or gunadira (indigenous names).

Photo: Almeida, F. C. M. 2013.

**Characteristics: there** are 174 species in Brazil, most do not exceed 100 g in body weight, communicate and orient themselves by echolocation, and have few cones in the retina. They also have a prominent nasal leaf and are sexually dimorphic.

**Habitat and distribution:** there are two basic types of habitat (refuges): diurnal and nocturnal. Daytime refuges are used most of the time and are the places where bats sleep and carry out their social activities. Nocturnal refuges serve as shelter during the bats' period of activity. It is a cosmopolitan order.

**Behavior:** the nasal leaf is important for directing the ultrasound coming out of the nostrils. During the echolocation process, they transmit high-frequency sounds through their mouths or noses which are reflected by surfaces in the environment.

**Feeding:** Most are frugivores - they eat a wide variety of fruits, such as mangoes, bananas, almonds, figs, papaya, guavas and especially wild fruits. It's very common to see them in cities, feeding on mango and almond trees. They are very important seed dispersers.

**Reproduction:** they are polyestrous, with an average of one offspring per year, and almost all species have only one pair of teats. They use sound to mate.

**Habits:** nocturnal.

**Importance:** they are considered harmful to fruit trees because they attack the fruit in orchards. However, they are biologically important for dispersing seeds, and economically: they serve as a model for the creation of various technologies, such as sonars, radars, ultrasound devices and even medicines for circulation.

## DID YOU KNOW?

*D* During fertilization, "delayed implantation" can occur, which is when zygotes do not implant promptly in the uterus after fertilization, but are stored temporarily until the organism allows the pregnancy to proceed at a more favorable time (depending on the food supply and temperature).

L Some Phyllostomidae common in Brazil can catch up to 500 insects per hour.

## PARDAL

KINGDOM: Animmalia

phylum: chordata

CLASS: Birds

ORDER: Passeriformes

SUBORDINATE: Passeri

PARVORDEM: Passerida

FAMiLY: Passaridae

GÈNERO: *Passer*

SPECIES: *Passer domesticus*

POPULAR NAME: sparrow, common sparrow.

Photo: Rodrigues, J. 2010.

**Characteristics:** males and females are sexually dimorphic, with the former characterized by a black bib, grey forehead and crown, dark lores and a brownish back with dark markings. Females have neither a bib nor dark lores, with brownish plumage and a cream stripe from the eye to the nape of the neck. The beak is thick, as is typical of granivorous birds.

**Habitat and distribution:** they are cosmopolitan birds and adapt well to urbanized areas and living with humans. It is currently the bird species with the widest geographical distribution. Its natural distribution is in Eurasia and North Africa

**Behavior:** they can form flocks. They can also behave aggressively and displace other birds trying to nest in their territory.

**Food:** varies depending on the season. Almost all year round it eats all the seeds it can find, but during the breeding season it looks for more insects.

**Reproduction:** the nest is spherical with a side entrance, made of grass, feathers, paper, cotton and other fibers, exceptionally made by the male. It is built in cavities and crevices far from the ground, in trees, roofs, streetlights and traffic lights. Other birds' nests can also be used. The 4 speckled gray eggs are incubated by the pair for 12 days. The chicks are fed small arthropods and leave the nest at around 10 days old, when they switch to a vegetarian diet. The chicks often return to the nest to sleep for a while.

**Habitat:** the sparrow is quite abundant throughout the country and is generally ubiquitous in humanized areas, both in large cities and in inhabited villages. It occurs all year round and can form large flocks, especially in agricultural areas or in urban park roosts.

**Importance:** displaces native species due to competition for environmental resources. This species is in the process of extinction in cities, as urban changes have driven flocks to the countryside. Scientists and experts say that serious changes in the urban ecosystem in recent times have had a huge impact on the sparrow population, whose numbers are steadily declining. They believe that the birds cannot return unless some measures are taken now.

## DID YOU KNOW?

o name sparrow means: domestic bird, bird that lives in houses.

It arrived in Brazil around 1903 (according to historical records), when the then mayor of Rio de Janeiro, Pereira Passos, authorized the release of this exotic bird from Portugal.

# FLUFFY TURTLEDOVE

KINGDOM: Animmalia

phylum: chordata

CLASS: Birds

ORDER: Columbiformes

FAMILY: Columbidae

GENRE: *Columbine*

SPECIES: *Columbina squammata*

POPULAR NAME: turtledove, firecracker, rattlesnake turtledove, paruru.

Photo: Pignaton, G. 2010.

**Characteristics: It** measures 19.5 cm in length and is also known as the rattlesnake (due to the noise it makes with its wings, which resembles a rattle). It has a beige background color, with light and dark spots that give it the appearance of scales. No sexual dimorphism.

**Habitat and distribution: it** inhabits fields, forest edges, savannahs, caatingas and wooded gardens in urban areas. Its distribution is in the Northeast and Central-West regions, in the states of Sao Paulo, Paranà and Rio Grande do Sul, as well as in French Guiana, Venezuela, Paraguay and Argentina.

**Behavior: it** roams in pairs or small groups along the edges of forests, savannahs, orchards, parks and other types of vegetation, excluding very open or very closed areas. Its silence is only broken by vocalization, which the bird only emits while perched in well-hidden places, and by the noise produced by the wings when the bird takes flight, resembling a moan.

**Food:** seeds, insects, worms and larvae.

**Breeding:** makes a nest of sticks in the shape of a cup, usually 1 or 2 m high, sometimes also on the ground. Lays 2 white eggs. Breeding period: June to November.

**Habits:** the firecrest is a dove with generally discreet habits, which lives in pairs or small groups. It feeds on the ground, walking with its belly almost touching the ground. When frightened, it flies abruptly to nearby trees.

**Importance:** this species is threatened by pollution, hunting and the destruction of its habitat.

### DID YOU KNOW?

The male and female take care of building the nest and the two eggs. They feed the chicks with a mixture of regurgitated seeds and a white liquid produced by glands in the esophagus.

This mixture is called "pigeon's milk" and is very rich in nutrients and is one of the reasons why the chicks leave the nest so quickly.

# WHAT'S THE DIFFERENCE?

## BETWEEN A BEE AND A WASP?

To the naked eye, bees can be distinguished from most wasps by the first segment of the thorax and the hind legs.

**Bees:**

+ The hairy chest;

+ The hind legs are wide and covered in pious-looking hairs, adapted to carrying pollen.

**Wasps**:

+ Chest with little hair;

+ The first segment of the thorax forms a collar around the neck that reaches the base of the forewings;

+ All paws are sturdy, thin and hairless.

## BETWEEN BUTTERFLIES AND MOTHS?

**Butterflies**:

+ Daytime habit.

+ When they land, they leave their wings raised.

+ Butterflies have thin antennae with an enlarged tip.

+ Butterfly caterpillars are generally hairless and their cocoons (chrysalis) look like leaves.

**Moths**:

+ Night owl (the majority).

+ When they land, they leave their wings open.

+ They have feather-like antennae (they have sensory bristles that serve to pick up pheromones from the opposite sex).

+ Their bodies are usually covered in hair.

+ Moth caterpillars usually have stinging hairs and make silk cocoons.

## ⅛⅛ BETWEEN CRICKET, GRASSHOPPER AND HOPE?

**Cricket**:

+ They are agile insects, small to medium in size - usually three centimetres.

+ They feed mainly on other insects and plants.

+ Crickets have antennae with more than 30 parts that are as long as their bodies.

+ The production of sounds (although not all crickets have the ability to sing). Only crickets with wings can make the cri cri cri. To produce the sound, you have to rub one wing against the other. Only males are capable of making this noise.

**Grasshopper:**

+ They are larger insects than crickets, usually measuring five to six centimeters in length.

+ Most feed on plants - herbivores.

+ They have short antennae compared to crickets.

+ Male grasshoppers also sing, but they produce the sound in a different way. Instead of rubbing their wings together, they rub their hind legs against them.

**Hope:**

+ They are usually larger than their relatives, measuring an average of eight centimetres.

+ Some species can feed exclusively on plants or other insects.

+ The antennae of hopefuls are even longer than those of crickets.

+ Most hopes are green.

+ They have wings that look like leaves. That way they can hide in a plant.

+ The sound of hope is the same as that of crickets and is produced in the same way: by rubbing one wing against the other.

# NAVIGATING...

j http : //www.biodiversidadeteresopolis.com. br/

j http : //www.biologando .com.br/

j http://borboletaskmariposas.blogspot.com.br/

j http://borboletasbr.blogspot.com.br/

j http://dimaserose.blogspot.com.br/

j http : //euamolouvadeus.wix.com/mantodea

j http://www.insetologia.com.br/

j http://www.insectaagbiologicos.com.br/

j http://www.mundobiologia.com/

j http://netnature.wordpress.com/

j http://tudolevaapericia.blogspot.com.br/

j- http://www.vivaterra.org.br/

## GLOSSARY

Aphids are tiny insects that feed on plant sap, also known as aphids.

**Appendix:** a protruding part of the body, or organ, of an animal with a variety of functions: movement, assimilation of nutrients, protection, among others.

**Cosmopolitan:** in biogeography, this is a group of living beings that can be found practically anywhere in the world.

**Detritivore:** an animal that feeds on organic waste.

Sexual **dimorphism** is considered when there are male and female individuals of a species with different physical characteristics.

**Entomophages:** animals that feed exclusively on insects.

Wing **span:** the greatest distance between the tips of the wings.

**Epigynum: this** is the external genital structure of female spiders, consisting of a hardened and slightly raised plate located in front of the epigastric groove, in which the openings of the seminal receptacles are located.

**Pheromones:** these are sex hormones that living beings release at a certain stage in their lives.

**Phytophages:** living beings that feed on plants.

**Phytophile:** one who lives on plants.

**Frugivores:** animals that feed on fruit.

**Fusiform:** spindle-shaped organism, i.e. elongated and with the ends narrower than the center.

**Gonopore:** genital opening of some invertebrates.

**Hematophagous: a** living creature that feeds on blood.

**Herbivore:** a living being that feeds on plants.

**Holometabolous:** an animal that undergoes complete metamorphosis during its development.

**Imago:** adult stage or reproductive stage of an insect

**Insectivore:** animal that feeds on insects.

**Lichens:** these are very simple living beings that constitute a symbiosis of an organism formed by a fungus and an alga or cyanobacterium.

**Mimicry: the** property that some animals have of taking on the color or configuration of objects in their environment, or of any being belonging to that environment.

**Notochord: a** semi-rigid bundle of cells surrounded by a fibrous sheath, between the dorsal nerve cord and the digestive tract, which supports the organism.

**Oostegites:** are structures that originate from the thighs of pereiopods.

**Ootheca:** is the name used in entomology for a kind of case formed by the secretion of certain insects and containing a cluster of eggs.

**Operculum:** eyelid or covering in a chamber-like structure.

**Ovopositor:** an organ used by arthropods to lay eggs.

**Parthenogenesis:** a biological process in which an egg develops into a new living being without fertilization.

**Pereiopod:** the legs themselves (crustaceans).

**Pleopod:** the paired abdominal appendages of crustaceans.

**Polyestrus:** animals that have two or three reproductive peaks during the year.

**Polyphagous:** animals that eat everything, both plant and animal foods.

**Ruderal:** is the name given in ecology to plant communities that develop in heavily disturbed environments.

**Seasonal:** is a characteristic of an event that always occurs at a certain time of year.

**Solanaceae:** a botanical family of angiospermic plants.

**Ubiquitous:** omnipresent; everywhere at once.

**Urticating:** cells or organs of plants or animals that cause irritation, itching to the touch or a burning sensation in the affected area.

# REFERENCES

## INVERTEBRATES

### SILVER SPIDER

CRASH, Cesar. **Argiope spider in Rio de Janeiro**. Insetologia - Insect Identification, 2012. Available at: <http://www.insetologia.com.br/2012/12/aranha-argiope-no-rio-de-janeiro.html> Accessed on: April 03, 2013.

GONZAGA, M. O.; SANTOS, A. J.; JAPYASSÙ, H. F. **Ecologia e comportamento de aranhas.** Rio de Janeiro: Interciência, 2007. 400p.

SILVA, A. da. **Feeding strategy of the garden spider, *Argiope argentata* (Araneae: Araneidae)**. Ebah. Available at: <http://www.ebah.com.br/content/ABAAAewycAK/modo-alimentaçãoo-aranha-jardim-argiope-argentata> Accessed on: May 23, 2013.

WIKIPEDIA. **Silver spider.** Wikipedia: the free encyclopedia, 2013. Available at: <http://pt.wikipedia.org/wiki/Aranha-de-prata> Accessed on: 22 Apr 2013.

### SPIDER FLYCATCHER

DANIEL. **Flycatcher spider (*Philaeus chrysops*)**. Wild Territory. 2011. Available at: <http://territorioselvagem.forumeiros.com/t48-aranha-papa-moscas-philaeus-chrysops> Accessed on: Feb. 20, 2014.

GONZAGA, M. O.; SANTOS, A. J.; JAPYASSÙ, H. F. **Ecologia e comportamento de aranhas.** Rio de Janeiro: Interciência, 2007. 400p.

VIVATERRA. **PAPA MOSCA (*Philaeus sp.*).** VIVATERRA - SOCIETY FOR ENVIRONMENTAL DEFENSE, RESEARCH AND EDUCATION. Rio de Janeiro. Available at: < http://www.vivaterra.org.br/aracnideos.htm#papamosca > Accessed on: Feb. 20, 2014.

### BICHO PAU

DUARTE, J. Jr. **ALBUM OF BRAZILIAN LIVING BEINGS:** Woodworm. EE José Duarte Júnior. 2012. Available at: <http://eejoseduartejr.blogspot.com.br/2012/02/album-de-seres- vivos-brasileiros.html> Accessed on: 12 Jan 2014.

PEREIRA, P. G. et al. **Insects in Order**. Porto Botanical Garden: Casa Andresen, 2012. p. 53-54.

SANTANA. A. L. **Insects: Woodworm**. Infoescola. Available at: <http://www.infoescola.com/insetos/bicho-pau/> Accessed on: 12 Jan. 2014.

VIVATERRA. **Insects:** BICHO PAU (*Bactridium grande*). VIVATERRA - SOCIETY FOR

ENVIRONMENTAL DEFENSE, RESEARCH AND EDUCATION. Rio de Janeiro. Available at: < http://www.vivaterra.org.br/insetos.htm#bichopau> Accessed on: January 12, 2014.

**SMALL TOMATO BORER**

BENVENGA, S. R. ***Neoleucinodes elegantalis*** **(Guenée) (LEP.: Crambidae) in staked tomato plants:** population dynamics, level of control with sex pheromone and efficiency of pesticides. 2009. 144 f. Thesis (PhD in Agronomy - Agricultural Entomology) - Faculty of Agricultural and Veterinary Sciences, Jaboticabal campus, São Paulo State University, São Paulo, 2009.

BIOCONTROL. ***Neoleucinodes elegantalis***. PESTS. Available at: <http://www.biocontrole.com.br/?area=pragas&id=12> Accessed on: 07 Jan 2014.

BUG, Biological Agents. **Parasitoids:** 1- *Trichogramma pretiosum* and *T. galloi.* BUGBrasil. Sao Paulo. Available at: <http://www.bugbrasil.com.br/parasitoides.asp> Accessed on: Feb. 10, 2014.

CRASH, Cesar. **Small Tomato Borer in Piaui**. INSETOLOGY - Insect Identification, 2014. Available at: < http://www.insetologia.com.br/2014/01/broca-pequena-do- tomateiro-no-piaui.html> Accessed on: 07 Jan 2014.

LORSBAN. **Small fruit borer:** *Neoleucinodes elegantalis.* Dow AgroSciences.

Available at: <http://www.dowagro.com/br/lorsban/pragas/brocapequena.htm> Accessed on: 07 Jan 2014.

**CRISOPA**

BANTZ, R. and STOLZEL, E. **Biology ofthe green lacewing *Chrysoperla carnea* and its use in biological control.** BOKU. 2012. Available at:

<http://www.dnw.boku.ac.at/fileadmin/_/H95/H953/Angewandte_Entomologie/WS11_12/Bantz_St%C3%B6lzel.pdf> Accessed: July 28, 2013.

INSETS. **Crisopa.** Getting to know the animals. 2012. Available at: <http://connhecendoosanimasdomundo.blogspot.com.br/> Accessed on: July 22, 2013.

LETARDI, A.; COELHO, J.; ANDRADE, R. ***Chrysoperla carnea*** **(Stephens, 1836).** NATURDATA. 2011. Available at: <http://naturdata.com/Chrysoperla-carnea-38350.htm> Accessed on: July 25, 2013.

*NUNEZ*, E. Ciclo biològico y crianza de *Chrysoperla externa* y *Ceraeochrysa cincta* (Neuropteta, Chrysopidae). **Revista Peruana de Entomologia.** Peru: 1988. vol. 31. p. 76-82

REYES, David Marquina. **Neuroptera: chrysopa.** MACREANDO. Spain, 2012. Available

at:<http://macreando.blogspot.com.br/search/label/Neuroptera#axzz2uBCzXGwG> Accessed: 23

Jul 2013.

WIKIPEDIA. **Chrysopidae.** Wikipedia: the free encyclopedia, 2013. Available at: <http://pt.wikipedia.org/wiki/Chrysopidae> Accessed on: July 28, 2013.

**CYANOPEPLA**

CYANOPEPLA. ***Cyanopeplajucunda***. Taxonomy. Available at: <http://lepidoptera.pro/taxonomy/88214> Accessed on: 04 Jan. 2014.

CYCLOPAEDIA. ***Cyanopepla jucunda***. Cyclopaedia.net. 2012. Available at: <http://sv.cyclopaedia.net/wiki/Cyanopepla_jucunda> Accessed on: 04 Jan. 2014.

FERRO, V. G.; DINIZ, I. R. Arctiidae (Insecta: Lepidoptera) from the Boracéia Biological Station (Salesópolis, Sao Paulo, Brazil). **Biota Neotrópica**. 2007. v.7. n° 3.p. 331-338.

GRUBE, W. ***Cyanopepla jucunda*** (Walker, 1854). Butterflies and moths. Paranà. Available at: <http://borboletaskmariposas.blogspot.com.br/search/label/Arctiidae> Accessed on: 04 Jan 2014.

LIMA, Costa. Lepidoptera: superfamily Noctuoidea. *In:* **Insects of Brazil** - National School of Agronomy. PARTE III - 6° TOMO. p. 160-245.

**EMBUA**

CONZO, H. **Snake lice - A bit about many legs**. Pest blog. 2009.

Available at: < http://www.blogdaspragas.com.br/2009/05/piolhos-de-cobra-um-pouco-sobre-muitas.html> Accessed on: 02 Mar. 2014.

NUTRIHUMUS. **Snake lice.** Nutrihumus- creation of earthworms, soils, substrates and fertilizers. 2013. Available at: <http://eucriominhocas.blogspot.com.br/> Accessed on: 02 Mar. 2014.

---

**HOPE**

EVARISTO, M. **Insects: hope**. Curiosity about animals, 2014. Available at: <http://curiosidadesanimais2013.blogspot.com.br/2014/01/insectos.html> Accessed on: Feb. 20, 2014.

PEREIRA, P. G. et al. **Insects in Order**. Porto Botanical Garden: Casa Andresen, 2012. p. 114

TREKNATURE. **Hope**. Treknature. 2007. Available at: <http://pt.treknature.com/gallery/photo129420.htm> Accessed on: Feb. 20, 2014.

---

**SPIDER CAT**

Toxicological Information Center - SC. **Animals - caterpillars**. Available at: < http://www.cit.sc.gov.br/index.php?p=lagartas> Accessed on: May 15, 2013.

MONTEIRO, R. F. et al. Composition, abundance and notes on the ecology of lepidopteran larval species associated with five host plant species in the Restinga de Jurubatiba National Park, RJ. **Brazilian Journal of Entomology**. Sao Paulo: 2007. vol.51 n°.4

WILSON, José. **Spider caterpillar** - *Phobetron hipparchia.* Behavioral entomology. 2010. Available at: <http://zehentomologus.blogspot.com.br/2010/06/lagarta-aranha-phobetron-hipparchia.html> May 15, 2013.

**PRAISE - GOD**

FAVACHO, César. **Main Families - Mantidae**. I love praying mantises. 2013. Available at: < http://euamolouvadeus.blogspot.com.br/2013/07/principais-familias-mantidae.html > Accessed on: 8 Jan 2014.

PEREIRA, P. G. et al. **Insects in Order**. Porto Botanical Garden: Casa Andresen, 2012. p. 36-38

**ANTS**

CRASH, Cesar. **Piaui ant bug.** INSETOLOGY - Insect Identification, 2014. Available at: < http://www.insetologia.com.br/2014/01/percevejo-formigao-no-piaui.html> Accessed on: 12 Jan 2014.

GÓMEZ, D. R. S. et al. **Manual de identificaçao de insetos e outros invertebrados da cultura da soja**. Londrina: Embrapa Informaçao Tecnológica, 2006. p. 66

MOREIRA, H. J. C.; ARAGÂO, F. D. **Manual de Pragas da Soja**. Campinas: 2009. p. 144.

**PIRATE BUG**

BASE de Datos de Invertebrados Introducidos a Galâpagos. ***Rasahus hamatus*** (F.). Species: show. Available at: <http://rockbugdesign.com/invert_ref/es/species/show/370/> Accessed on: 03 Mar. 2014.

CRASH, Cesar. **Pirate bug of Minas Gerais**. INSETOLOGY - Insect Identification, 2014. Available at: <http://www.insetologia.com.br/2014/01/percevejo-pirata-em-minas- gerais.html> Accessed on: 07 Jan 2014.

LIMA, Costa. Hemiptera: superfamily. Peiratinae. In: **Insects of Brazil** - National School of Agronomy: 1940. PART III - 2ND VOLUME. Chap. XXII p. 212-216.

VIQUIPÈDIA. **Redùvid.** 2014. Available at: <http://ca.wikipedia.org/wiki/Red%C3%BAvid> Accessed on: 03 Mar. 2014.

**SERRA PAU**

CRASH, Cesar. **coleoptera cerambycidae (Longicorns or Serra-paus).** Insectology - Insect identification. Available at: <http://www.insetologia.com.br/search/label/Coleoptera%20Cerambycidae%20%28Longic%C3%B3rnios%20or%20Serra-paus%29> Accessed: 22 Jul 2013.

ENTOMOSERVICE. **cERAMBYcIDAE:** Cerambycinae. Entomoservice: Insects of collection. Available at: <http://www.entomoservice.fr/index.php?cPath=30_36&sort=2a&page=5> Accessed on: 20 Jul 2013.

LIMA, Costa. Coleoptera: family Cerambycidae. *In:* **Insetos do Brasil - PARTE** III - 9° TOMO. Escola Nacional de Agronomia:1955, n° 11. 67- 143 p.

MAX. **Chainsaw**. Welcome to the world of animals: Sao Paulo, 2012. Available at: <http://animalgt.blogspot.com.br/2012/08/serra-pau.html> Accessed on: 27 Jul 2013

ZORZENON, F. J.; CAMPOS, A. E. de C.; JÙNIOR, J. J.; POTENZA, M. R. Main pests in urban forestry III. *In: Boring* **Insects.** Sao Paulo: Health R&D Center V, no. 171, 2012.

**CLAY WASP**

CAMILLO, E. **The natural history of the mud-dauber wasp *Sceliphron fistularium* (Hymenoptera: Sphecidae) in southeastern Brazil**. NCBI, 2002. Available at: <http://www.ncbi.nlm.nih.gov/pubmed/12298237> Accessed on: 12 Jan 2014.

CRASH, Cesar. **Amassa-Barro wasp in Minas Gerais**. Insectology - Insect identification, 2013. Available at: <http://www.insetologia.com.br/search/label/Hymenoptera%20%28sem%20taxon%29%20Vespas%20e%20Marimbondos> Accessed on: 12 Jan 2014.

FONSECA, V. L. I.; OLIVEIRA, M. A. C. **Insects that visit flowers:** how to recognize and observe them. ProCiències Project - FAPESP / CAPES 2000. Available at: <http://www.webbee.org.br/didatico/pg02_ch_vespa.htm> Accessed on: 12 Jan 2014.

PEREIRA, P. G. et al. **Insects in Order**. Porto Botanical Garden: Casa Andresen, 2012. p. 114

---

## VERTEBRATES

**ANU**

CAMPOS, D. **Black Anu (*Crotophaga ani*).** Passarinhando.2008. Available at: < http://dimaserose.blogspot.com.br/2008/07/preto-crotophaga-ani.html> Accessed on: 10 Dec. 2013.

VIVATERRA. **Birds: anu**. VIVATERRA - SOCIETY FOR DEFENSE, RESEARCH AND ENVIRONMENTAL EDUCATION. Rio de Janeiro. Available at: <http://www.vivaterra.org.br/aves.htm#anu> Accessed on: 10 Dec. 2013.

**IGUANA**

ARAGUAIA, M. **Iguana (Familia Iguanidae).** Brazil School. Available at: <http://www.brasilescola.com/animais/iguana.htm> Accessed on: 02 Mar. 2014.

BARTEN, S. L. **Breeding and management of the Green Iguana**. Translation of the article "Bayer Exotics Symposium - Selected Papers on the Green Iguana and Antimicrobials. In Exotic Pets - pages 13-22. Bayervet. Available at: <http://www.zoopets.com.br/iguana/apostila.pdf> Accessed on: 02 Mar. 2014.

CAMPOS, Z. **Estudo das Iguanas no Pantanal.** Mato Grosso do Sul: ADM/Embrapa Pantanal, 2004. n° 67.p. 1-2.

Green IGUANA changes color to protect itself. **Diàrio do Grande ABC**. 2012. Available at: <http://www.dgabc.com.br/Noticia/21887/iguana-verde-muda-de-cor-para-se-proteger> Accessed on: 02 Mar. 2014.

SILVA, M. M.; CÂNDIDO, P. R.; SOUZA, C. R. **Study on Iguanas (*Iguana iguana*) existing in the woods of the Centro Universitàrio Luterano do Brasil Ji-Paranà.** Rondônia. 2006.

**MORCEGO**

GOMES, M. N.; UIEDA, W. Diurnal shelters, colony composition, sexual dimorphism and reproduction of the hematophagous bat *Desmodus rotundus* (E. Geoffroy) (Chiroptera, Phyllostomidae) in the State of Sao Paulo, Brazil**. Brazilian Journal of Zoology**. Curitiba: 2004. vol.21. n°.3

MELLO. M. **House of Bats**. SBEQ. Available at: <http://www.casadosmorcegos.org/casadosmorcegos/Sobre_pt.html> Accessed on: 5 Mar 2014.

REIS, Nélio R. et al. **Bats of Brazil**. Londrina: 2007. p.253.

**FIRE-BREATHING TURTLEDOVE**

BIRDS of Brazil. **Fogo-apagou, rolinha-cascavel**. Domestic birds. Available at: < http://www.avedomestica.com.br/cocad/index.php?option=com_content&view=article&id=2791:fogo-apagou-rolinha-cascavel&catid=182:avesdobrasil&Itemid=65> Accessed on: 12 Dec. 2013.

MARTINS, C. **Fofo-apagou**. WikiAves. 2009. Available at: <http://www.wikiaves.com.br/fogo-

apagou> Accessed on: 12 Dec. 2013.

PIGNATON. G. **Fogo- apagou.** Gerê's Blog. 2010. Available at:

<http://gereblog.blogspot.com.br/2010/12/fogo-apagou.html> Accessed on: 12 Dec. 2013.

VIVATERRA. **Birds: fire-extinguished**. VIVATERRA - DEFENSE, RESEARCH AND ENVIRONMENTAL EDUCATION. Rio de Janeiro. Available at:

<http://www.vivaterra.org.br/aves_2.htm#fogoapagou> Accessed on: 12 Dec. 2013.

**PARDAL**

LEÂO, T. C. C. et al. **Invasive alien species in the Northeast of Brazil:** contextualization, management and public policies. Recife: CEPAN- Instituto Horus, 2011. p. 101.

RODRIGUES, J. **Common sparrow:** *Passer domesticus.* Portugal: Avesdeportugal.info. 2010. Available at:< http://www.avesdeportugal.info/pasdom.html> Accessed on: 10 Mar. 2014

WIKIAVES. **Sparrow**. Wikiaves. Available at: < http://www.wikiaves.com.br/pardal> Accessed on: 12 Mar. 2014.

**SUPPORT REFERENCE:**

APRILE, M. **Butterflies and moths:** Characteristics and differences between the two. Biology. Education-UOL. 2006. Available at:

<http://educacao.uol.com.br/disciplinas/biologia/borboletas-e-mariposas-caracteristicas-e-diferenças-entre-as-duas.htm> Accessed on: 06 May 2013.

CIT. **Animals - caterpillars**. Toxicological Information Center. Santa Catarina. Available at: <http://www.cit.sc.gov.br/index.php?p=lagartas> Accessed on: May 9, 2013.

D'ARAÙJO e SILVA, A. G.; C. R. Gonçalves; D. M. Galvao; A. J. L. Gonçalves; J. Gomes; M. N. Silva & L. Simoni. 1968. **Quarto Catàlogo dos Insetos que Vivem nas Plantas do Brasil, seus Parasitos e Predadores** - ParteII - 1°Tomo**.** Rio de Janeiro, IBGE Foundation, 622 p.

LEÂO, T. C. C. et al. **Invasive alien species in the Northeast of Brazil:** contextualization, management and public policies. Recife: CEPAN- Instituto Horus, 2011. p. 101.

PEREIRA, P. G. et al. **Insects in Order**. Porto Botanical Garden: Casa Andresen, 2012. p. 114

POUGH, F. H.; JANIS, C. M.; HEISER, J. B. **The life of vertebrates**. Sao Paulo: Atheneu, 2008. ed. 4ª . p. 684

RUPPERT, E.E., FOX, R.S. and BARNES, R.D. **Zoology of Invertebrates**. Sao Paulo: Ed. Roca,

2005, 7th ed. 1145 p.

SOARES. L. G. S.; PEREIRA, M. R. **What's the difference between cricket, grasshopper and hope?**

Traveling in the world of science. 2011. Available at:

<http://viajandonomundodaciencia.blogspot.com.br/2011/08/qual-diferenca-entre-grilos-gafanhotos.html> accessed on: 03 Mar 2014.

SOARES, R. R. **Morphology and Physiology of Vertebrates I**. Teresina: UFPI/UAPI, 2009. p.146

MIX
Papier aus verantwortungsvollen Quellen
Paper from responsible sources
FSC® C105338

Printed by Books on Demand GmbH, Norderstedt / Germany